Débora Luise Rocha de Carvalho
Kristerson Reinaldo de Luna Freire
Adna Cristina Barbosa de Sousa

# CULTIVATION OF WHITE SHIMEJI (PLEUROTUS OSTREATUS VAR. FLORIDA)

Débora Luise Rocha de Carvalho
Kristerson Reinaldo de Luna Freire
Adna Cristina Barbosa de Sousa

# CULTIVATION OF WHITE SHIMEJI (PLEUROTUS OSTREATUS VAR. FLORIDA)

## Axenic Cultivation of White Shimeji- (Pleurotus Ostreatus Var. Florida) on Regional Agro-Industrial Co-Products

ScienciaScripts

Cover image: www.ingimage.com

This book is a translation from the original published under ISBN 978-613-9-68475-5.

Publisher:
Sciencia Scripts
is a trademark of
Dodo Books Indian Ocean Ltd. and OmniScriptum S.R.L publishing group

120 High Road, East Finchley, London, N2 9ED, United Kingdom
Str. Armeneasca 28/1, office 1, Chisinau MD-2012, Republic of Moldova, Europe
Printed at: see last page
**ISBN: 978-620-8-31330-2**

***CULTIVATION OF WHITE SHIMEJI (PLEUROTUS OSTREATUS VAR. FLORIDA)***

***AXENIC CULTIVATION OF WHITE SHIMEJI (PLEUROTUS OSTREATUS VAR.***

***FLORIDA) IN REGIONAL AGRO-INDUSTRIAL CO-PRODUCTS***

***DÉBORA LUISE ROCHA DE CARVALHO[1]***

***FEDERAL UNIVERSITY OF PARAÍBA CAMPUS I/BIOTECHNOLOGY CENTRE -***

***CBIOTEC)***

***DR. KRISTERSON REINALDO DE LUNA FREIRE [2]***

***PROFESSOR AT THE FEDERAL UNIVERSITY OF PARAÍBA CAMPUS I/CENTRE OF***

***BIOTECHNOLOGY - CBIOTEC***

***DR[A] . ADNA CRISTINA BARBOSA DE SOUSA[3]***

***PROFESSOR AT THE FEDERAL UNIVERSITY OF PARAÍBA CAMPUS I/CENTRE OF***

***BIOTECHNOLOGY - CBIOTEC E-MAIL CONTACT: ADNACSOUSA@GMAIL.COM***

***JOÃO PESSOA - PARAÍBA JUNE - 2024***

## ACKNOWLEDGEMENTS

*I would first like to thank my mother Carla, for giving me the wings and support to be able to leave home and go to university 2,479 kilometres away. As a teacher, she always taught me the importance of education and science. To my brother Pedro, who was always with me and taught me to see fun in every situation. To my grandmother Luzia, for being my greatest example of strength and faith. My sister-in-law Andressa, whom I thank for joining our family and proving to be an incredible mum. And my dear niece Mavie, who has brought so much happiness into our lives and made me happy with her learning. To all of you, thank you for the support and strength you have given me on this journey, even from far away I think of you every day.*

*I would also like to thank my supervisor Adna Cristina, who is like a second mother to all of us in her laboratory, always there to help and advise, and an example of a professional and human being. Together with the entire team at Cbiotec's Molecular Genetics and Plant Biotechnology Laboratory, this work is the fruit of everyone who has passed through Adna's hands. I would also like to thank Professor Kristerson for making some of the materials used in this research available. I would especially like to thank my lab partner, Lipe, for making everything more fun, less tiring and simple, and for being my support beyond those walls, always listening to me and looking for solutions to my problems, as well as sharing his music with me while we did the experiments.*

*Finally, I'd like to thank my friends, Lipe, Medeiros, Pedro, Anderson, Aninha, Malu, Karine, Rian and Andrea, who made these years so much lighter with all our games between classes, our suppers in geographies and outings to de-stress from the routine, these years wouldn't have been the same without you. And especially Emille, my friend since primary school, who shared a flat with me and was always by my side all these years, from Altamira to João Pessoa.*

## SUMMARY

*Pleurotus ostreatus var. florida, known as white shimeji, is one of the most widely consumed mushrooms in the world and in Portugal. This species is of great interest due to its sensory and nutritional aspects, as well as the fact that it is cheaper to grow and requires less specific conditions. Currently, there is a growing population scenario, which increases the demand for food and products to meet the needs of the population, generating a greater production of co-products. As a result, there is a need to reuse these materials, and mushroom cultivation has become a biotechnological strategy for valorising and reusing these by-products. The aim of this study was to assess the feasibility of reusing agro-industrial by-products generated from bean cultivation and craft beer production, analysing the carbon and nitrogen contents present in the formulations tested, correlating them with biological and production parameters, as well as analysing the ideal C:N ratios for mycelial growth and basidiom development. Four formulations were tested: T1: Grain-free green bean pods (80%) + malt bagasse (20%); T2: Grain-free green bean pods (80%) + malt bagasse (20%) + residual yeast (5%); T3: Grain-free green bean pods (80%) + malt bagasse (20%) + hot trub (10%); T4: Grain-free green bean pods (80%) + malt bagasse (20%) + hop trub (10%). In this study, it was seen that the addition of hop trub (T3) and hot trub (T4) increased the carbon content and reduced the nitrogen content, which resulted in higher C:N ratios, but supplementation with residual yeast did not significantly affect the carbon and nitrogen content, as it had similar results to T1, which did not contain any brewery liquid waste. As for the mycelial run rate, there were no significant differences between the formulations, with T1, T2 and T4 completing the run in 24 days, while T3 completed it in 25 days, showing that the C:N ratio range of the treatments tested did not influence mycelial growth. In terms of production parameters, the treatment supplemented with hop trub (T4) had the best performance in terms of yield (19.5%) and biological efficiency (76.8%), and was also the treatment with the highest C:N ratio (39:1). The humidity of the basidiomata ranged from 81 to 83 per cent. The rates of organic matter loss and compost consumption between the 4 treatments show that it is possible to obtain more than one productive flow. It can be concluded that all the carbon/nitrogen ratios obtained varied within a suitable range for this crop, so T1, T3 and T4 proved to be ideal for P. ostreatus var. florida production.*

***Keywords:*** *Mushrooms; lignocellulosic waste; Pleurotus spp.*

**SUMMARY**

# 1. INTRODUCTION

Fungi have been known since ancient times for their medicinal and edible properties. The fungal kingdom comprises more than a million species, including mushrooms, which are of great ecological importance due to their ability to degrade organic matter. Currently, edible mushrooms have gained prominence due to their high nutritional value and bioactive compounds that offer benefits when added to the human diet, as well as being a versatile ingredient with great potential in the food industry (GONZÁLEZ, 2020).

An important example are those belonging to the Pleurotus genus, Pleurotus ostreatus var. florida being a species popularly called the white shimeji- or oyster mushroom because it has a basidiome shaped like an oyster. One of the most notable characteristics of this genus is its ability to grow on different types of substrates. This is due to its ability to degrade various materials, thanks to the presence of specific enzymes. This property is extremely important when considering the application of these fungi in the utilisation of agro-industrial co-products/by-products, also known as waste (RITOTA; MANZI, 2019).

With population growth, the number of debates on food security is growing, as the need to produce food is increasing in order to guarantee access to food and adequate nutrition for the world's population. In order to meet this growing demand, there has been a rise in the industrial sector, which has led to an increase in the production of by-products from this area, such as agro-industrial co-products that are often disposed of inappropriately in the environment, such as disposal in landfills or burning, increasing pollution rates and causing various environmental problems (KUMLA, 2020).

Therefore, there is a need to reuse these by-products in order to reduce the negative effects generated by the agro-industry. One alternative is through the cultivation of edible mushrooms using the by-products as substrates for the development of the fungus. One of the techniques used is axenic cultivation, since the substrate undergoes sterilisation processes to guarantee the growth of the fungus of interest, avoiding possible contamination that could be present in the formulation made up of the waste, which maximises this production.

These by-products include those from bean production, such as bean stalks after the beans have been threshed, and those from the production of craft beer which generate a large amount of solid waste, such as malt bagasse, and liquid by-products such as residual yeast, hot trub and hop trub, which are the result of the coagulation of high molecular weight proteins and leftover hops. Both (bean pods and malt bagasse) are present and important crops in the northeast region, as well as being lignocellulosic materials, i.e. their composition is made up mainly of cellulose, hemicellulose and lignin. Species of the Pleurotus genus are heterotrophic decomposer organisms and therefore possess a series of specific enzymes capable of degrading lignocellulosic materials, such as those present in these coproducts.

Another important aspect for this cultivation is understanding the carbon-nitrogen (C:N) ratio, since this proportion plays a fundamental role in the mycelial race and subsequent formation of the basidiome of the edible mushroom, especially for the Pleurotus genus. Having a good C:N ratio is key to optimising production and ensuring successful cultivation (BELLETTINI, 2019).

The cultivation of these fungi can therefore be used as a biotechnological strategy to reduce and valorise agro-industrial by-products. Contributing to reducing the environmental impact caused by these materials, as well as exploring concepts of a sustainable and circular economy, where these by-products are reused and valorised. It also guarantees the production of foods with a high nutritional value, as is the case with edible mushrooms. Based on this, the aim of this work was to verify the effect of the C:N ratio on the mycelial growth and production of P. ostreatus var. florida using regional by-products as a substrate.

# 2. OBJECTIVES

## 2.1 OBJECTIVE GENERAL

Check the effect of the C:N ratio on the mycelial growth and production of P. Ostreatus var. florida using regional by-products as substrate

## 2.2 OBJECTIVES SPECIFIC

To evaluate the potential for using bean pods without the grain and malt bagasse supplemented with hot trub, hop trub and residual yeast in the formulation of a growing medium for mushroom production;
Characterise the biological parameters: substrate colonisation time, primordia emission period, basidiom formation time, basidiom diameter, pileus height and pileus width;
Characterise the production parameters: yield, biological efficiency, productivity, compost consumption and loss of organic matter of the different formulations;
Carry out the chemical characterisation of the formulations in terms of carbon (C), nitrogen (N) and C:N ratio;
Classify the best formulation for vegetative development and production of fungal biomass.

## 3. THEORETICAL BACKGROUND

### 3.1. Mushroom cultivation in Brazil and the World

Mushrooms are fungi that are highly valued for their nutritional, medicinal, therapeutic and functional properties. They have been known and used by mankind since the dawn of civilisation, especially on the Asian continent, where they are considered important components of traditional medicine in these countries.

In ancient times, it was commonly believed that mushrooms had magical properties, such as prolonging youth, giving strength to soldiers, and were used by many cultures in rituals and on special occasions. In Ancient Egypt, they were considered plants of immortality and were therefore reserved for Egyptian royalty (AHMED et al., 2023). Some studies also discuss the ancient use of psychoactive mushrooms to achieve metaphysical and spiritual experiences; in Mayan culture, these mushrooms were also present and were described as food for the gods (KOTOWSKI, 2019).

Nowadays, edible mushrooms are appreciated for their health benefits, such as their nutritional value, as they are an excellent source of proteins, vitamins (B1, B2, B12, C, D and E), minerals, fibre, carbohydrates and have a low lipid content. They also contain compounds such as unsaturated fatty acids (MAJESTY, 2019), phenolic compounds, ascorbic acid and carotenoids, making them ideal from a nutraceutical point of view (GUPTA et al., 2018). For this reason, they are sought after to complete the diets of the population seeking a healthier life, as they provide all these nutrients and have a reduced calorie content.

In the global context, China is the largest producer and consumer of mushrooms. This leadership is attributed to an ancient cultural tradition and the rapid growth of this industry on the Asian continent, driven by the low cost of production and the high profitability of mushrooms (RAMAN, 2020). According to Urben (2017), per capita consumption in China is 8.0kg/year, compared to 4.0kg/year in Germany, 2.0kg/year in France, 1.3kg/year in Italy and just 0.288kg/year in Brazil.

Although Brazilian consumption of mushrooms is lower than in other countries, the most consumed mushrooms in Brazil are Agaricus bisporus (Champignon de Paris), species of the Pleurotus genus, mainly P. ostreatus (Shiimeji) and Lentinula edodes (Shiitake), with production of 9,000, 8,500 and 5,000 tonnes per year, respectively, as stated by the Secretariat of Agriculture, Livestock and Rural Development - RS (2020). According to data from the Ministry of

Industry, Foreign Trade and Services (MDIC), there was a 70% increase in the import of pickled mushrooms, from around 7,000 tonnes to more than 12,000 tonnes between 2015 and 2022, which indicates an opportunity for growth in mushroom consumption by the Brazilian population (STEFFEN et al., 2019).

We can also mention the search for mushrooms as an alternative to animal protein-based foods, since it is a protein product with around 20-37% protein in its dry weight centesimal composition of mushrooms of the A. bisporus and P. ostreatus species (PAZZA, 2020). It is important to note that there is an increase in the number of vegetarians and vegans in Brazil, as shown by a survey carried out by the Brazilian Vegetarian Society conducted by Ibope (2018), which found that almost 30 million people declared themselves vegetarians, which represents 14% of the national population, and shows that by 2023 the expectation is that this number could exceed 40 million, representing 20% of the national population (Brazilian Vegetarian Society, 2021).

As for cultivation, China has become a major producer of mushrooms mainly because of the development of the "Jun-Cao" cultivation technique, which translates as "mushrooms" (Jun) and "grasses" (Cao), which was created in 1983 and has conferred social, ecological and economic benefits on the cultivation of edible and medicinal mushrooms (FERREIRA,2020). This technique is a sustainable and innovative approach. The method consists of using grasses instead of wood and sawdust as a substrate for cultivation. It has promoted the high efficiency and quality of Chinese mushrooms, and in 1990 the country became the world's largest producer and is still doing so today (URBEN, 2017).

Another cultivation method is by solid-state fermentation, in which the fungi grow on solid substrates without the presence of free water. The substrate widely used for cultivation in Asia is rice straw and cotton waste, which are cheaper and more suitable for mushroom cultivation. This cotton waste has attracted Asian farmers and encourages this method of cultivation (RAMAN, 2021).

In Brazil, cultivation methods are the result of Asian colonisation since they were introduced in the 1950s in the Mogi das Cruzes region of São Paulo, and this region is currently responsible for the majority of mushroom production in the country (LACERDA, 2021). Because this production was introduced by the Chinese, the technology adopted in the country is the Jun-Cao technique with some modifications, using national and regional substrates (URBEN, 2017). Generally, substrates made from sugar cane bagasse are used. sugar, soya straw, oat straw, corn cobs, wheat straw, rice straw, grasses, fruit pulp and peel, banana leaves and various types of fodder (BERNARDI; NASCIMENTO, 2020).

However, there is still a need to increase cultivation techniques in the Brazilian scenario in order to make these fungi a food accessible to the entire population, as it is an important alternative to combat malnutrition, considering its high nutritional quality (FERREIRA, 2020).

## 3.2. Biological aspects and nutritional characteristics of Pleurotus ostreatus

P. ostreatus is one of the most widely produced edible mushroom species in the world and in Brazil, due to its nutritional quality and the presence of compounds of medicinal importance, as well as its taste and texture, which are appreciated by the population. They are known as white rot fungi because they are considered excellent lignocellulosic decomposers due to their high mycelial growth rate and enzymatic capacity. These characteristics favour their cultivation, as they can quickly colonise various substrates such as agronomic waste (ZÁTERE-SALAZAR, 2020). Mushrooms of this genus are known for their great morphological variety, presenting different colours, shapes and sizes, as shown in Figure 1.

**Figure 1:** Morphological variations in the basidioma formed by Pleurotus spp.

A - Pleurotus purpurea-olivaceus. B- Pleurotus djamor. C- Pleurotus ostreatus. D- Pleurotus citrinopileatus. E- Pleurotus giganteus. F- Pleurotus ostreatus var. florida. Source: Adapted from GALAPPATHTHI et al. (2021); ZENG et al. (2022).

The morphology of mushrooms is one of the ways to differentiate and classify them into different categories, such as the different orders. Mushrooms belonging to the order Agaricales have a basidiocarp with two obvious parts: the stipe or stem and the pileus or cap (Figure 2). Another important structure is the hymenium, which contains the lamellae. These are structures whose function is to harbour the basidiospores or sexual spores, so this is the fertile part of the mushroom (UBERN, 2017).

**Figure 2:** Mushroom morphology.

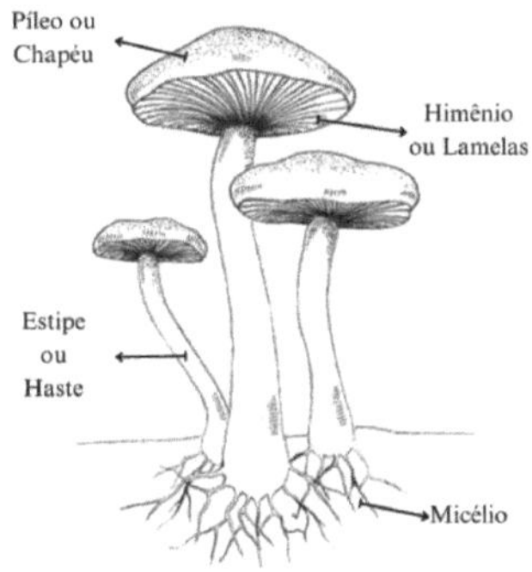


Source: Adapted from Roberto Suárez's drawing, available at https://museudaamazonia.org.br/fungos-2/

P. ostreatus var. florida, popularly known as White Shimeji-, belongs to the kingdom Fungi, phylum Basidiomycota (called basidiomycetes), class Agaricomycetes, order Agaricales, family Pleurotaceae and genus Pleurotus (SILVA, 2024). As shown in Figure 3, its main characteristics are the presence of a pileus in the shape of a shell, spatula or oyster, which is why it can be known as an oyster mushroom. short, lateral and off-centre stipe that vary in length and thickness, they have an elongated, ellipsoid and cylindrical spore, and this variation has a white to creamy colour (JARIAL, 2020).

**Figure 3:** Morphology of Pleurotus ostreatus var. florida.

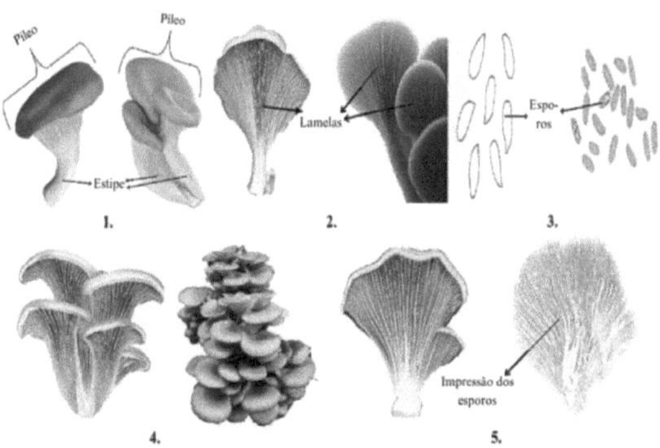


1. Pileus and stipule; 2. lamellae; 3. spores; 4. fruiting body; 5. spore impression. Source: Adapted from Jarial et al., 2024.

The life cycle of mushrooms can be divided into three phases involving the formation of a mycelium (vegetative part), basidiome (fruiting body) and basidiospores, which are the sexualised spores. According to Tortora (2016), basidiomycetes are teleomorphs because they have a meiosporic stage, i.e. they are able to produce sexual spores and survive asexually through the development of the vegetative part (Figure 4).The mycelium (vegetative part) is made up of a set of hyphae, which grow by elongating their ends. If a fragment of the hypha is broken, it generates a new propagule that can continue to grow and develop to form a new mycelium. Hyphae that contain cell walls called septa are also called septate hyphae, where the nuclei separate. In a few classes of fungi, the hyphae do not have this separation and are called hi s no iti s (TORTORA, 2016). Therefore, when a new organism emerges through the fragmentation of hyphae, or the formation and dissemination of asexual spores, asexual reproduction occurs, since there is no interaction between two organisms and there is no genetic variability. Sexual reproduction in mushrooms is mediated by basidiospores (sexual spores) which are contained in the lamellae above the basidia. When mature, they are released into the environment. They germinate and give rise to new hyphae and the cycle repeats itself (MANAN, 2021). In this sexual process, two nuclei are joined together by the fusion of hyphae, called plasmogamy, resulting in dikaryotic hyphae, i.e. hyphae with two nuclei. These hyphae have indefinite growth, are septate and have an important structure, the connecting staples, which connect one hypha to another, binding them together. Subsequently, these nuclei fuse in the process of karyogamy, generating sexual spores that will be released into the environment (URBEN et al., 2017; MANAN, 2021).

**Figure 4:** Reproductive cycle of basidiomycetes.

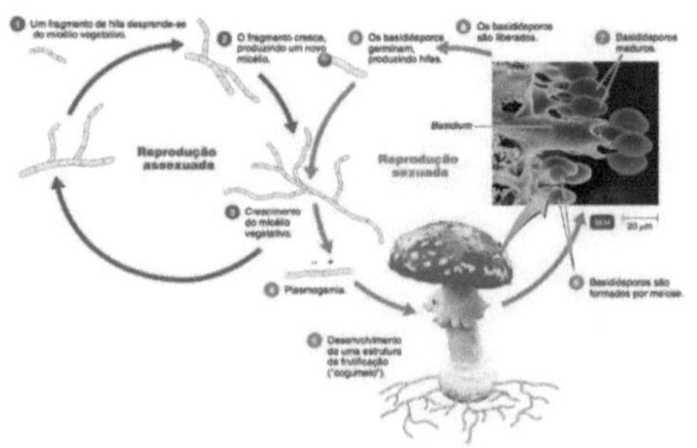

Source: TORTORA, 2016.

The different species of the Pleurotus genus are important for their high nutritional value, with a higher protein content compared to vegetables. However, it is lower compared to meat and milk. On the other hand, they contain all the essential amino acids. They also contain minerals such as potassium, iron, phosphorus, calcium and vitamins (RAMAN, 2022).
Another essential point is that they have a variety of active compounds that bring numerous benefits to humans, such as immunomodulatory, anti-parasitic, anti-cancer, antioxidant, anti-inflammatory, antibacterial and anti-hyperglycaemic properties. Some of the compounds responsible for this series of properties are: hydroxylated triterpenes, lectin, alpha-glucans, glucans, phenolic compounds, carotenoids and various polysaccharides (SOCCOL et al., 2017; GIACOMETI, 2020; REIS JUNIOR et al., 2021).

### 3.3. Axenic cultivation of mushrooms

Naturally, fungi propagate by spores (sexual or asexual), which, when they find a suitable environment, colonise the substrate, propagating by division, forming hyphae that give rise to mycelia, forming a new organism that can be the same as the matrix or with variability if there is sexual reproduction. In artificial cultivation, one of the methods adopted is axenic cultivation, in which the substrate is prepared at the start of the process and sterilised, allowing the fungus to grow without competition from other microorganisms and achieve greater productivity in organic substrates (OLIVEIRA, 2020).
The stages of this method are shown in Figure 5 and consist of: (1) Preparation of the inoculum, which is also called seed or spawn. This initial part involves inoculating a primary mycelium from a matrix into a previously treated grain, so that vegetative growth occurs for subsequent inoculation into the desired substrate; (2) Preparation of the substrate, which must contain the nutrients and conditions necessary for the growth and development of the fungus, such as carbon, nitrogen and the correct humidity. This stage includes mechanical processes such as cutting and grinding the substrate if necessary, in order to facilitate the colonisation of the fungus to be inoculated. As it is an axenic culture, it is at this stage that the substrate is sterilised to prevent the growth of other microorganisms and consequently contamination of the substrate; (3) Inoculation, which consists of placing the fungus in contact with the substrate, following the established protocol to guarantee colonisation without contamination of the medium; (4) Incubation, which is the period of growth of the fungus in the substrate until it reaches total colonisation and finally; (5) Fruiting, comprising the period of growth of the primordia until the point of

harvest, where the fruiting bodies are fully formed (LACERDA, 2021; SILVA, 2024).

**Figure 5:** Stages of axenic cultivation.

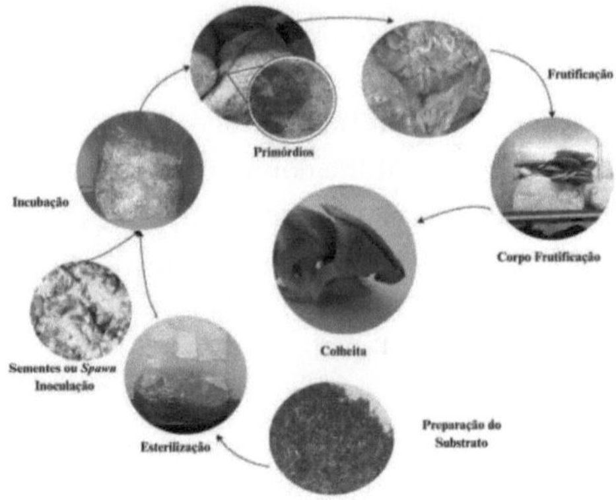


Source: Adapted from Silva et al. 2024.

### 3.4. Co-products agro-industrial

Agro-industrial co-products are those obtained from agricultural biomass through industrial processing, which can be physical (milling and cutting), chemical (esterification), thermochemical (combustion) and biochemical (fermentation). These by-products include liquids generated from chemical inputs and biofuels, gases, solid waste (micropollutants such as microplastics) and lignocellulosic solids with a high organic matter content from processing in the food and feed sector (VAZ JUNIOR, 2020). P. ostreatus var. florida is of great biotechnological interest because it plays an important role in the recycling and degradation of materials, mainly because it has lignocellulosic enzymes capable of degrading agro-industrial by-products, making it an economically viable, sustainable process that adds value to the end product.

These enzymes are divided into oxidative enzymes such as lignases, lacases and manganese peroxidases, and hydrolytic enzymes such as cellulases, xylanases and tanases. It is due to the presence of these enzymes that fungi are able to utilise lignocellulosic substrates, making it possible to reuse agro-industrial by-products (BELLETTINI, 2019).

### 3.4.1. Crop of beans

Beans belong to the Fabaceae family (legumes), the Phaseolus vulgaris L species, and are a highly valued food by Brazilians, as they are an excellent source of energy and contain low fat content (COELHO, 2020). Brazil can produce up to three crops a year, making it the world's third largest producer of this crop (CONAB, 2019).According to data from the Brazilian Institute of Geography and Statistics (IBGE) (2022), the bean production sector produced 2,842,395 tonnes with a harvested area of 2,607,616 hectares. According to the National Supply Company (Conab), bean production for the 2023/24 season is expected to average around 3.3 million tonnes, 7.6% more than the 2022 harvest. This result is the sum of production from the first and second harvests, together with the field survey carried out in June 2024, plus forecasts for the third harvest. In the Northeast, around 1.2 million hectares are used to grow beans, as shown in Figure 6. This crop occupies the third largest planting area in the region, behind soya and maize, which emphasises the importance of beans in local agricultural production.

**Figure 6:** Data on the area planted or destined to be harvested in the Northeast region.

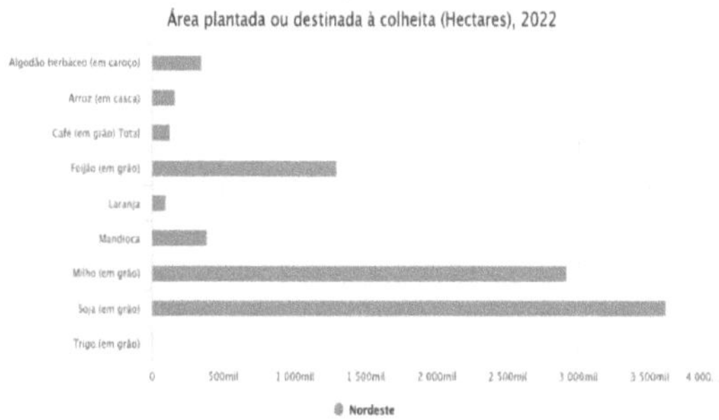


Source: IBGE - Municipal Agricultural Production, 2022.

This high production generates large quantities of by-products, mainly the bean pods after threshing the beans (Figure 7), which are discarded. It is possible to reuse these biological materials, as they are essentially lignocellulosic compounds, i.e. they are made up of lignocellulose, including cellulose, lignin and hemicellulose (NINDJIO, 2022).

**Figure 7:** Green bean pod.

Source: EMBRAPA Image Bank, 2022.

According to Egídio (2019), the lignocellulosic materials in bean pods contain an average of 41% cellulose, 20.6% hemicellulose and 5.3% lignin. In terms of elemental analysis, it has approximately 40.65% carbon and 1.90% nitrogen (DIONIZIO, 2019). Due to this composition, the bean pod can be reused as a substrate for mushroom production, since bean cultivation currently does not have a sustainable destination for the pod, and it contains essential components for fungiculture.

### 3.4.2. Brewing

Beer is a fermented alcoholic drink made from four main ingredients: water, malt, hops and yeast. According to MAPA (Ministry of Agriculture and Livestock), Brazil is the third largest producer of beer in the world, behind China and the United States. In 2023, Brazil's trade balance in the beer market recorded the highest result of the study period, totalling US$ 147,191,235.

Although the Northeast region ranks third in terms of the number of registered breweries, with 122 establishments, it ranks second in terms of declared production volume, with 3.3 billion litres per establishment, which represents 21.9% of the national volume (MAPA, 2023). According to Amoriello (2021), for every 1,000 tonnes of beer produced, around 10,000 tonnes of liquid by-products and 137-173 tonnes of solid waste are generated in the form of waste water, malt bagasse, residual yeast, residual hops, rootlets and unwanted material. It is therefore necessary to utilise these co-products so that they are not disposed of inappropriately. Barley, Hordeum vulgare L., is one of the main ingredients in brewing beer. This grain goes through a malting process before being added to production. The beverage production process begins with the milling of the malt, followed by the saccharification and filtration stage of the wort, generating bagasse after this process, which is the main resulting solid waste, accounting for around 85% of brewing waste (Figure 8). The other by-products are: the trub, which corresponds to the fraction of coagulated proteins and hops in the process, which is divided into hot and hop trub, and the residual yeast, which are the microorganisms used to ferment the drink (MATHIAS et al., 2014; COELHO NETO et al., 2020).

**Figure 8:** Malt bagasse.

Source: Dos Santos Giuliani et al. (2019)

Malt bagasse is composed of 29.92% hemicellulose (made up of xylan and arabinan), 20.80% total lignin (insoluble and soluble lignin) and 15.99% cellulose, all of which are considered fibres, resulting in an average fibre content of 65-70%. This material is also rich in protein at 21.16%, as well as having 8.33% extractives and 3.76% ash. It therefore has essential nutrients for the growth and development of mushrooms (Table 1) (MASSARDI, 2020).

**Table 1:** Chemical composition of malt bagasse

| Componente | Teor (%) |
|---|---|
| Celulose | 15,99 ± 0,88 |
| Hemicelulose, sendo: | 29,92 ± 1,60 |
| Xilana | 25,22 ± 1,46 |
| Arabinana | 4,71 ± 0,14 |
| Lignina, sendo: | 20,80 ± 0,42 |
| Lignina insolúvel | 17,78 ± 0,41 |
| Lignina solúvel | 3,02 ± 0,01 |
| Proteinas | 21,16 ± 0,61 |
| Extrativos | 8,33 ± 0,76 |
| Cinzas | 3,76 ± 0,05 |

Source: MASSARDI, 2020.

The yeast used in the brewing process is Saccharomyces cerevisiae. Despite being in an inactivated state when discarded, its composition is rich in proteins, minerals, fibres and vitamins. It is made up of 25-35% non-cellulosic carbohydrates, 17-25% cellulose, 15-24% protein, 8-28% lignin, 10% lipids and 5% ash (KARLOVIé et al., 2020). Hot trub is a by-product of beer rich in proteins, fibres and sugars, derived from coagulated proteins and the remains of hops and brewer's wort added during boiling, with around 23.2% proteins, 26.0% fibres, 22.9% starch, 17.3% sugars, 5% fat and 3.3% ash (LOMUSCIO et al., 2022). Hop trub is the result of the purging of solid material from the bottom of the fermenter after the cold hopping or dry hopping stage, which is mostly made up of hops, 43% cellulose and lignin, 15-30% total resins (a- B-acids), 15% protein, 10% water, 8% ash, 4% polyphenols, 2% monosaccharides and pectins, as well as essential oils, monosaccharides and steroids (DURALOO;

SILVA; BOGUSZ, 2019). It also contains beer and probably yeast in smaller quantities. Therefore, residual yeast and trubs can be reused for mushroom cultivation, acting as sources of nitrogen supplementation, as both contain a significant amount of protein in their composition, as well as being carbon sources due to the presence of high fibre contents.

### 3.5. Substrate composition and the impact of the carbon/nitrogen ratio

The growing medium is the main factor for good mushroom production, which must provide the vital nutrients for mycelial and fruiting body growth and development, since the composition of this substrate is one of the critical factors that will influence these stages positively or negatively.

One of the factors that most interferes with mushroom cultivation is the carbon/nitrogen ratio (C:N), which must be in constant balance depending on the stage of production; this factor influences mycelial growth, yield and production efficiency (SALAME, 2016).

Carbon is the most abundant element in the substrate and is the main component of lignocellulosic waste. It acts as a primary energy source in the synthesis of important organic compounds and is vital for fungal metabolism. It is present in the form of various organic compounds such as cellulose (KUMLA et al., 2020).

Nitrogen also plays an essential role in the metabolism of these fungi, as it is converted into amino acids, nitrogenous bases and proteins. For this reason, some crops/substrates receive nitrogen supplements in order to produce mushrooms with a higher protein value, since substrates with a lower nitrogen concentration generate mushrooms with a lower protein content (MUSWATI et al., 2021).

The different species of the Pleurotus genus grow best when supplemented with organic nitrogen sources such as amino acids (asparagine, alanine, glycine, arginine and tryptophan), along with ammonium and yeast extract, rather than mineral nitrogen (MLECZEK, 2021; SILVA, 2024). However, a high level of nitrogen can suppress the production of lignocellulolytic enzymes, which affects the degradation of lignin, preventing the development of the mycelium and subsequent formation of the fruiting body (OKERE et al., 2021).

According to Berllettine (2019), higher proportions favour mycelial growth, resulting in a shorter time to complete colonisation of the substrate. While lower ratios favour the formation of the fruiting body, increasing productivity. It is therefore necessary to have an ideal range of C:N ratios depending on what is required. In the Pleurotus genus, the ideal C/N ratio is between 20 and 50:1 (ALBUQUERQUE; SOUSA, 2021).

# 4. MATERIAL AND METHODS

The experiments were conducted at the Laboratory of Molecular Genetics and Plant Biotechnology (LGMBiotec), Biotechnology Centre, Federal University of Paraíba (UFPB), Campus I, João Pessoa, Paraíba.

## 4.1. Origin of Mushroom

The matrix of P. ostreatus var. florida belongs to the fungal culture collection of LGM Biotec/CBIOTEC/UFPB and is being maintained on 2 % Sabouraud-dextrose agar medium (5 g/L meat peptone, 20 g/L glucose, 5 g/L casein peptone and 15 g/L bacteriological agar) (Figure 9), according to the methodology described by Furlan et al. (1997).

**Figure 9:** Primary matrix of Pleurotus ostreatus var. florida.

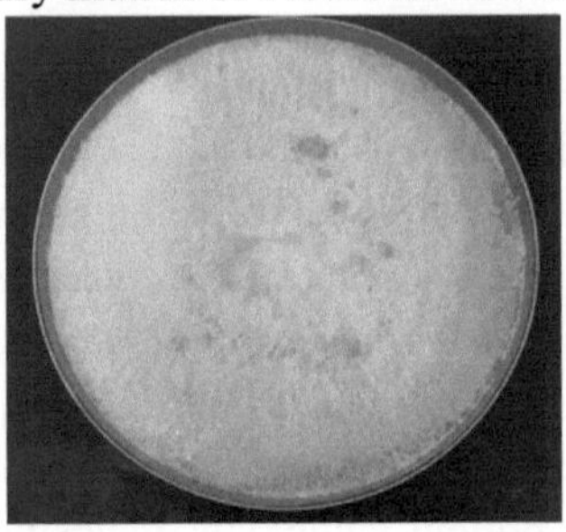

Source: Own authorship (2024).

## 4.2. Substrates and supplements used

Popcorn maize (Zea mays var. everta), malt bagasse (Hordeum vulgare L.) and green bean pods without the grain (Phaseolus vulgaris) were used as substrates. Hop trub, hot trub and residual yeast (S L g r™ W- 34/70) were used as supplements (Figures 10 and 11).

**Figure 10:** Substrates used.

A- Popcorn corn (Zea mays var. everta), B- Green bean pods without the grain (Phaseolus vulgaris) and C- Malt pomace (Hordeum vulgare L). Source: Author (2024).

**Figure 11:** Brewery liquid waste used as supplements.

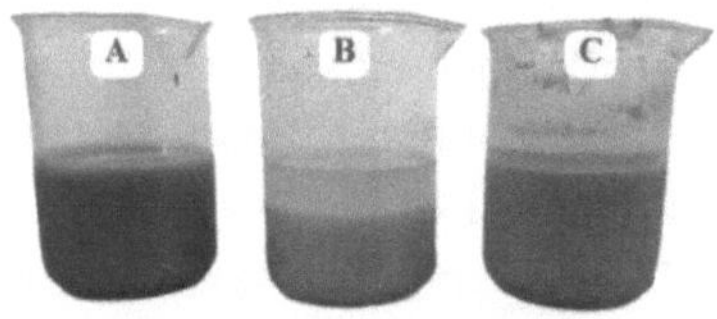


A- Residual yeast, B- Hot trub and C- Hop trub.
Source: Own authorship (2024).

### 4.3. Origin of substrates

The beer by-products (malt pomace, hop trub, hot trub and residual yeast) were supplied by a craft brewery in the city of João Pessoa. The green bean pods were purchased from a street market in the metropolitan region of João Pessoa/PB. The popcorn used was of commercial origin.

### 4.4. Processing of Substrates

**Popcorn:** the kernels were cooked over high heat in a pressure cooker for 90 minutes. After cooking, the extract from the cooking process was drained and the kernels were left at room temperature (28 ± 2°C) to dry externally for approximately 60 minutes. They were then packed in plastic bags and kept refrigerated at 4°C (BONATTI et al., 2004).

**Malt bagasse: this** did not receive any specific treatment, and due to its humidity, no distilled water was added during preparation. After collection, the pomace was packed in plastic bags and kept refrigerated at 4°C.

**Bean pods without the grain:** initially they were dehydrated at 70ºC for 72 hours in a drying oven. Next, it was submerged in tap water for 24 hours and, after draining the water, it was placed on a flat platform for 2 hours to dry. Finally, the pods were ground in a knife mill and packed in plastic bags and kept refrigerated at 4°C until use.

**Residual yeast, hot trub and hop trub:** did not receive any particular processing. After collection, they were packed in plastic buckets and kept refrigerated at 4ºC.

### 4.5. Seed production

Initially, the processed popcorn was placed in a glass jar (6.0 x 8.5 cm) filling 80% of its volume. The jar was then closed with aluminium foil and autoclaved at 121ºC for 15 minutes. After cooling to room temperature (28° ± 2°C), three discs of mycelium with a diameter of 5 ± 1 mm were inoculated into the flask. Incubation took place at 28° ± 2°C, in the absence of light, until the surface was completely colonised (Figure 12).

**Figure 12:** Preparation of Pleurotus ostreatus var. florida seed.

A- Boiled maize after sterilisation and B- Substrate after colonisation. Source: Author (2024).

### 4.6. Qualitative assessment of growth mycelial

The substrates were weighed and, where necessary, mixed with their respective supplements. The final weight was 60g, 48g of bean pods (80%) and 12g of malt (20%), with 5µl/g 1 v ur 10 µl/g the hot and hop trub (Table 2). The treatments were placed in bioreactors made up of plastic polythene bags (2.5 x 15 cm), partially closed with polyamide ties and autoclaved at 121°C for 60 minutes (Figure 13). After 24 hours, 3 corn kernels with adhered mycelium were added to each bioreactor. The treatments were incubated at 28° ± 2°C in the absence of light for 15 days. Longitudinal growth was measured with a millimetre ruler (cm) to obtain the average mycelial growth of each formulation. The total colonisation time of each treatment was assessed, taking into account the time needed for all the specimens in the triplicate to reach complete colonisation of the surface. Popcorn was used as a control. To assess the vigour/density of the mycelium during the mycelial run, the treatments were compared qualitatively.

**Table 2**: Formulations for growing Pleurotus ostreatus var. florida.

| Treatments | | | Formulations and proportions | | | | | |
|---|---|---|---|---|---|---|---|---|
| T1 | Pods | of | Beans without beans | | | - | (80 | %) |
| + Malt bagasse - (20 %) | | | | | | | | |
| T2 | Pods of bean | | | without | grains | - | (80 | %) |
| | + Bagasse of | | | malt | - | | (20 | %) |
| + residual yeast - (5 %) | | | | | | | | |
| T3- Pod of pod without beans -(80%)<br>+ Malt bagasse - (20 %)<br>+ hot trub - (10 %) | | | | | | | | |
| T4- Podof pod without beans-(80%) | | | | | | | | |
| + Malt pomace - (20 %) + Hop trub - (10 %) | | | | | | | | |

**Source:** Own authorship, 2024.

**Figure 13:** Formulations for growing Pleurotus ostreatus var. florida.

**Formulations: TC:** maize (control); **T1:** Grain-free green bean pods (80%) + malt bagasse (20%); **T2:** Grain-free green bean pods (80%) + malt bagasse (20%) + residual yeast (5%); **T3:** Grain-free green bean pods (80%) + malt bagasse (20%) + hot trub (10%); **T4:** Grain-free green bean pods (80%) + malt bagasse (20%) + hop trub (10%). Source: Author (2024).

### 4.7. Formulations for the production of Pleurotus ostreatus var. florida.

For this experiment, 100g of substrates T1, T2, T3 and T4 were weighed and, where necessary, mixed with their respective supplements (Table 2). The treatments were placed in bioreactors made up of polythene plastic bags (12 x 20 cm), partially closed with polyamide ties and autoclaved at 121°C for 60 minutes (Figure 14). After 24 hours, 3 corn kernels with adhered mycelium were added to each bioreactor. The treatments were incubated at 28° ± 2°C in the absence of light for 15 days. The experiment was carried out in quintuplicate. After incubation, the bioreactors were opened to assess fruiting potential. The fruiting bodies were harvested when showed maximum development, as seen by the beginning of the unfurling of the edges of the pileus.

**Figure 14:** Formulation for the production of Pleurotus ostreatus var. florida.

**T1:** Green bean pods without grains (80%) + malt bagasse (20%); **T2:** Green bean pods without grains (80%) + malt bagasse (20%) + residual yeast (5%); **T3:** Green bean pods without grains (80%) + malt bagasse (20%) + hot trub (10%); **T4:** Green bean pods without grains (80%) + malt bagasse (20%) + cold trub (10%). Source: Authors, 2024.

### 4.8. Evaluation of biological parameters

During the cultivation period, biological aspects of the mushroom were analysed, such as: time taken to colonise the substrate, period of primordia emission, time taken to form basidiomata, height and width of the stipe (cm) and diameter of the pileus (cm).

### 4.9. Evaluation of production parameters

Yield, biological efficiency, loss of organic matter, compost consumption, moisture and productivity were assessed.

**Yield:** measured in terms of the fresh weight (g) of mushrooms produced per bag containing 100g of compost (Zakil et al., 2022).

**Biological efficiency (EB%):** the biological efficiency (EB%) of the process was determined by the ratio between the mass of fresh fruiting bodies and the mass of substrate based on their dry weight according to Equation 01 (Muswati et al., 2021).

$$EB(\%) = \frac{MBS}{MSS} \times 100 \quad \text{Eq.1}$$

MBS = Mass of fresh basidiocarps, g MSS = Dry mass of substrates, g

**Productivity:** the productivity (Pr g/day$^{-1}$ ) of the process was determined according to HOLTZ (2009). It consists of the ratio between the mass of dried fruiting bodies and the total cultivation time (time from inoculation to production flow, as shown in Equation 02.

$$Pr(g/dia) = \frac{MCS}{TC} \quad \text{Eq.2}$$

Where: MCS = Dry mass of fruiting bodies, g; TC = Total cultivation time, days.

**Loss of organic matter: an** index that assesses the decomposition of the substrate by the fungus. This index is based on the loss of organic matter decomposed by the fungus, which is determined by the difference between the dry mass of the initial substrate and the dry mass of the residual substrate (Equation 03) (RAJARATHNAM; BANO, 1989).

$$PMO\ (\%) = \frac{(MSSI - MSSR) \times 100}{MSSI} \quad \text{Eq. 03}$$

Where:

PMO= Organic matter loss, % MSSR= Residual substrate dry mass, g MSSI= Initial substrate dry mass, g

**Compost consumption percentages**: this index was calculated based on the dry mass of the substrate after cultivation divided by the initial dry mass of the compost, as shown in Equation 04.

$$CC(\%) = \frac{CF \times 100}{CI} \quad \text{Eq. 04}$$

Where: CF = Final dry matter compost; CI = Initial dry matter compost.

**Moisture content of the treatments and mushrooms:** this was determined by the gravimetric method using heat, which is based on the weight loss of the material when heated until it reaches a constant weight. All the treatments were kept in a drying oven at 105°C for 24 hours and the moisture content of the substrate was determined by the difference between the wet and dry biomass (Equation 05).

$$U(\%) = \frac{(MI - MF) \times 100}{MI} \quad \text{Eq. 05}$$

Where: MI = Initial sample mass; MF = Final sample mass.

### 4.10. N, C and C:N content

**Organic carbon:** determined using the Walkley and Black method, modified from Tedesco et al. (1995). Each sample (0.01 g) was added to an erlenmeyer flask containing potassium dichromate (K2Cr2O7) a i o sul üri o ($H_2$ $SO_4$ ) and heated to 60 °C. Two types of blank were prepared: one that was subjected to the same heating conditions and one that was not heated. After cooling, indicators were added and then the solution was titrated with ammoniacal ferrous sulphate (Fe (NH )$_{42}$ (SO )$_{42}$ .6$H_2$ O). The results were obtained from Equations 06 and 07:

$$A = \frac{((Vba - Vbn)(Vbn - Vba))}{Vbn + (Vba - Vam)} \quad \text{Eq. 06}$$

In which:
Vba: volume spent titrating the blank with heating;

Vbn: volume spent titrating the blank without heating; Vam: volume spent titrating the sample.

$$Carbono\ (C)\% = \frac{(A)(molaridade\ do\ sulfato\ ferroso)(3)\ x\ 100}{(massa\ da\ amostra\ (mg))} \quad \text{Eq. 07}$$

In which:
3= ratio between the number of moles of dichromate reacting with iron, multiplied by the number of moles of dichromate reacting with carbon, multiplied by the atomic mass of carbon.

**Total nitrogen:** the Kjeldahl method was used to analyse total nitrogen, where the samples (0.1g) were digested in a catalyst mixture containing sulphuric acid ($H_2$ $SO_4$ ) in a Velp Scientifica digester block until a blue colour was obtained. After cooling, the reactions were distilled in a Velp Scientifica nitrogen distiller in a basic medium (NaOH 35%) and the product of this distillation was collected in an erlenmeyer flask in a 4% boric acid solution ($H_3BO_3$) with indicators and titrated with 0.1 N hydrochloric acid (HCl) (Tedesco et al., 1995, Moretto et al., 2002). The nitrogen concentration was determined according to Equation 08:

$$N(N)\% = \frac{V \times Fc \times 0{,}0014}{M} \quad \text{Eq. 08}$$

In which:
V = volume of HCl used in the titration, mL. Fc= HCl correction factor
M = sample mass (g)

**C:N ratio:** The C/N ratio was obtained from the proportion of carbon contained in each sample in relation to nitrogen.

### 4.11. Statistical analysis

The experimental design will be entirely randomised complete blocks, using the Tukey test, with a significance level of $p<0.05$ (GOMES, 1990). The statistical analyses will be carried out using SAS (Statistical Analysis System) developed by SAS Institute Incorporation. The results were analysed using Tukey's test with a 5% significance level (ANOVA).

# 5. RESULTS AND DISCUSSION

## 5.1. Biological aspects of Pleurotus ostreatus var. florida

The life cycle of mushrooms can be divided into two phases: vegetative or asexual, consisting of the vigorous mycelium and the gathering of hyphae, forming rhizomorphic structures, and reproductive or sexual, consisting of the basidiom (fruiting body) and reproductive propagules such as spores. When this cycle is reproduced artificially there are a series of stages followed as shown in Figure 15. The first stage is to obtain a primary matrix of the mushroom on agar-Sabouraud-dextrose culture medium in order to preserve the culture and analyse the vegetative aspects. It is possible to visualise abundant growth of the fungus with a high-density, whitish cottony appearance (Figure 15-1). The main advantage of this vegetative propagation is that the mushrooms produced in this way have the same characteristics as the mushroom from which the material was taken. Thus, characteristics such as productive potential, fruiting body characteristics such as colour and shape, and other characteristics such as temperature sensitivity are maintained (ZHENG et al., 2021). In the primary matrix, it was possible to identify hyphae (Figure 15-2A) responsible for extracting nutrients from the substrates; connection clamps (15-2B) which are structures responsible for joining the hyphae, forming dikaryotic hyphae; and septa (15-2C), since P. ostreatus var. florida has septate hyphae, with a separation between their nuclei.The spawn, also called inoculum, inoculant or seeds (15-3), was obtained from the primary matrix. It was obtained by transferring small portions of mycelium from the primary matrix to flasks containing grain-type substrate (popcorn). Dense, vigorous and uncontaminated growth was observed. This stage is crucial, as a healthy and vigorous seed guarantees good initial development in the production substrate until it is completely colonised (Figure 15-4). The reproductive phase is characterised by the formation of macroscopic structures such as the fruiting body or basidiome (Figure 15-6A), in which the hymenium (Figure 15-5A) can be seen, located on the underside of the pileus or oyster-shaped hat, where the basidiospores are housed. The stem or stipe (15-5B), which is short, lateral and off-centre, and also the base or stalk (15-5C) of the mushroom were observed. The sexual spores (basidiospores) are usually elongated, cylindrical or ellipsoid (15-7). All these aspects visualised in this study are characteristic of this species (JARIEL, 2020).

**Figure 15**: Pleurotus ostreatus var. florida reproductive cycle.

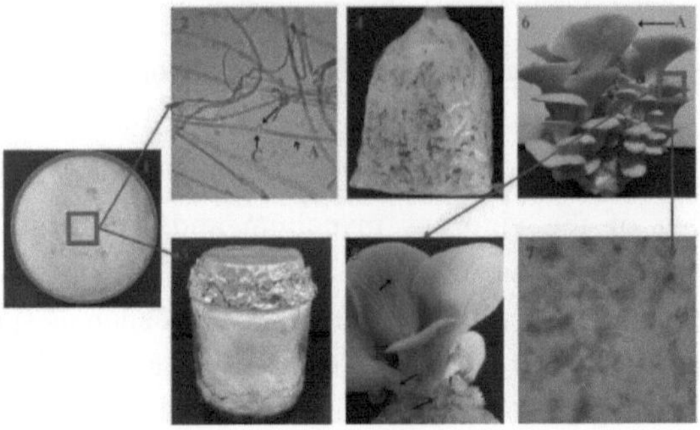

1: Colony of Pleurotus ostreatus var. florida with 15 days of growth on dextrose-Sabouraud agar; 2: Vegetative appearance: A- hyphae, B - connecting staple and C-septum; 3: Seed production on pre-cooked maize kernels; 4: Vegetative growth in production bags 15 days after inoculation; 5: Fruiting: 5A- Lamella-shaped hymenium, 5B-stem and 5C-base or foot (T3); 6: Fruiting: 6A- Hat or pileus and 7: Basidiospores (sexual spores). Source: Author (2024)

## 5.2. Evaluation of the mycelial development of Pleurotus ostreatus var. florida

Mycelial (vegetative) development was assessed by aspects of density, vigour, total colonisation of the substrate and growth time in the different formulations, since slow mycelium growth rates are an important problem in the mushroom production sector, affecting it negatively (KRUPODOROVA et al. 2024). Rapid colonisation of the substrate is necessary, as this will reduce the rates of contamination and rotting of the medium (SILVA, 2024).According to Table 3, the control treatment (TC) showed the shortest time for total colonisation of the substrate, with mycelial growth of 0.84 cm/day. While treatments T1, T2 and T4 completed colonisation in 24 days, with growth of 0.58, 0.57 and 0.62 cm/day, respectively. Treatment T3 took 25 days to complete colonisation, with growth of 0.58 cm/day.These results show that the different substrate formulations were able to provide the necessary nutrients for the vegetative growth of the mushroom, with a growth rate in line with other studies, such as those described by Silva et al. (2020) and Muswati et al. (2021), who described the time to complete the mycelial race as between 15-25 days.

**Table 3:** Evaluation of mycelial development based on the density and vigour of the different treatments for the production of Pleurotus ostreatus var. florida.

| Treatments | Total mycelial development time (days) | Mycelial growth (cm)• | Mycelial growth (cm/day) | U%* |
|---|---|---|---|---|
| TC (control) | 15,00 b | 12,60 | 0,84 a | 57 |
| T1 | 24,00 a | 8,75 | 0,58 a | 75 |
| T2 | 24,00 a | 8,50 | 0,56 a | 74 |
| T3 | 25,00 a | 8,76 | 0,58 a | 75 |
| T4 | 24,00 a | 9,30 | 0,62 a | 75 |
| CV%§ | 3,49 | 4,75 | 1,94 | |

Averages followed by the same letter do not differ by Duncan's multiple range test (P<0.05). Data not transformed.* Humidity of treatments. • Average growth up to the 15th day. § Coefficient of variation. Source: Author (2024).

Another parameter assessed was the humidity of the different substrates. All the treatments registered rates of between 74% and 75%, except for TC, which showed 57%. However, all the values are within the range considered ideal for fungal growth, which covers a humidity range of around 50% to 80% for fungal growth (BELLENTTINI et al. 2019; RATHOD et al., 2021). According to Silva et al. (2024), high substrate humidity can have negative effects on mycelium growth, as it can limit oxygen and reduce porosity, as well as increasing the chances of other organisms growing in the medium. Regarding the macroscopic aspect of the colonisation of Pleurotus ostreatus var. florida in the different treatments (Figure 16), it was possible to observe that the mycelium showed high vigour and density, with a white colour and compacted shape, which proves that the formulations could also be used for seed production (spawn).

**Figure 16:** Colonisation of Pleurotus ostreatus var. florida in the different treatments 15 days after inoculation.

**TC:** maize (control); **T1:** green bean pods without grains (80%) + malt bagasse (20%); **T2:** green bean pods without grains (80%) + malt bagasse (20%) + residual yeast (5%); **T3:** green bean pods without grains (80%) + malt bagasse (20%) + hot trub (10%); **T4:** green bean pods without grains (80%) + malt bagasse (20%) + hop trub (10%). Source: Author (2024).

### 5.3. Chemical characterisation of the formulations: carbon (C), nitrogen (N) content and C:N ratio.

The composition of the substrate is a critical factor that positively or negatively influences mushroom production. The carbon/nitrogen ratio (C:N) is especially important, and it must be balanced according to the production phases, as it can modify biological and production parameters.Carbon is the most abundant element in mushroom cultivation substrates and is the main component of lignocellulosic coproducts. It serves as the main source of energy for the proper functioning of fungal metabolism. In addition, nitrogen is also essential for successful cultivation, as the fungus needs an adequate amount of this element for the formation of its fundamental cellular structures, such as proteins and nucleic acids (KUMLA et al., 2020).P. ostreatus var. florida can acquire these essential elements from the substrate through the degradation of lignocellulose, cellulose, hemicellulose, lignin, starch and proteins. To do this, it uses degrading enzymes such as lacase, xylanase, cellulase, amylase and protease, which provide the substances for subsequent absorption by the mycelia. This process is closely associated with good mycelial growth, yield and productivity in mushroom cultivation (HU et al., 2023).According to Kumla et al. (2020), these enzymes are divided into: hydrolytic (cellulases and hemicellulases) are responsible for the degradation of polysaccharides; and oxidative (ligninases), which have the function of modifying and degrading lignin. Table 4 shows the carbon and nitrogen content and the carbon/nitrogen ratio of the formulations. It can be seen that T4 (supplemented with hop trub) had the highest carbon content (57.5%) and the lowest nitrogen content (1.80%), and therefore the highest C:N ratio (31.9:1). On the other hand, T1 (without supplementation, just the bean pod with the malt) and T2 (supplemented with residual yeast) achieved similar results, with carbon contents of 56.8% and 56.4% respectively, the nitrogen content was identical at 2.40% and the C:N ratio was 23.6:1 and 23.5:1. Treatment T3 (supplemented with hot trub) had a ratio of 29.9:1, with a carbon content of 56.4% and a nitrogen content of 1.90%.Therefore, it can be seen that supplementation with residual yeast did not significantly affect the carbon and nitrogen contents, as the results were similar to T1, which did not contain any brewery waste. However, the addition of hot and hop trub increased the carbon content and reduced the nitrogen content, which resulted in higher C:N ratios. This may have been the result of its high carbon content from fibres and starch, as it contains 26.0% fibres and 22.9% starch (LOMUSCIO et al., 2022).

**Table 4:** Carbon and nitrogen content and carbon/nitrogen ratio of the formulations.

| Treatments | C (%) | N (%) | C:N |
|---|---|---|---|
| T1 | 56,8 | 2,40 | 23,6: 1 |
| T2 | 56,4 | 2,40 | 23,5: 1 |
| T3 | 56,9 | 1,90 | 29,9: 1 |
| T4 | 57,5 | 1,80 | 31,9: 1 |

C(%): Carbon; N(%): Nitrogen; C:N: Carbon to Nitrogen ratio. **T1:** Green bean pod without the bean (80%) + Malt bagasse (20%); **T2:** Green bean pod without the bean (80%) + Malt bagasse (20%) + residual yeast (5%); **T3:** Green bean pod without the bean (80%) + Malt bagasse (20%) + hot trub (10%); **T4:** Green bean pod without the bean (80%) + Malt bagasse (20%) + cold trub (10%). Source: Author (2024).

As for the mycelial growth rate, there were no significant differences between the formulations, since T1, T2 and T4 completed the run in 24 days, while T3 completed it in 25 days. Therefore, in this experiment the C:N ratios obtained did not influence mycelial growth. However, according to Tavarwisa et al.

(2021), higher C:N ratios confer a higher mycelial growth rate, i.e. less time to complete the mycelial run.

### 5.4. Characterisation of the biological and production parameters of Pleurotus ostreatus var.florida

Based on the results of the mycelial growth experiments described above, all four treatments were selected for evaluation of biological and production parameters. Figure 17 shows the treatments fully colonised by the primary mycelium of Pleurotus ostreatus var. florida after 17 days of inoculation.

**Figure 17:** Substrates colonised by the primary mycelium of Pleurotus ostreatus var. Florida in the first stage of production.

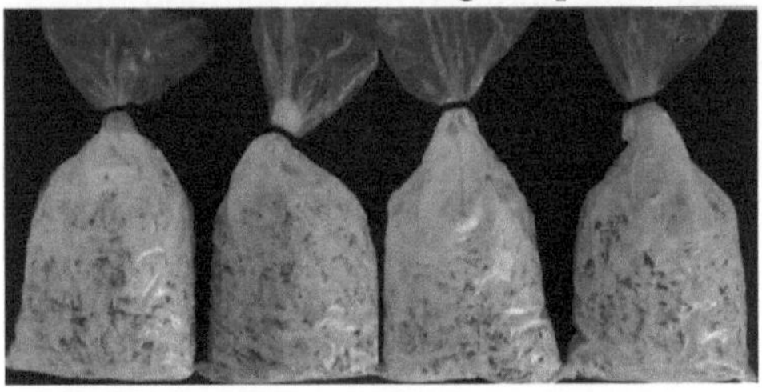

**T1:** Green bean pods without the grain (80%) + malt bagasse (20%); **T2:** Green bean pods without the grain (80%) + malt bagasse (20%) + yeast (5%); **T3:** Green bean pods without the grain (80%) + malt bagasse (20%) + hot trub (10%); **T4:** Green bean pods without the grain (80%) + malt bagasse (20%) + hop trub (10%). Source: Author (2024).

The characterisation of the biological aspects showed a total production cycle of 24 days for all the formulations tested, with 17 days for vegetative growth and 6 days for complete development of the basidiome (fruiting body) (Table 4). These results characterise an excellent cycle time, as these data were better than those found by Fufa et al. (2021) and Taskirawati et al. (2020), who describe a range of 28-45 days from inoculation to harvest for mushrooms of this genus. With regard to the size of the macroscopic structures, the length of the stipe varied from 1.40 to 1.80 cm. Treatment T1 had the longest stipe at 1.80 cm, followed by T4 with 1.70 cm, T2 with 1.60 cm and T3 with the shortest at 1.40 cm. Stipe width was also analysed and ranged from 0.63 to 0.90 cm. Treatment T2 had the highest value (0.90 cm), while T4 had the lowest. Treatments T1 and T3 had widths of 0.76 cm and 0.70 cm, respectively. Data obtained by Johnie et al. (2023) revealed similar results, with the width of the stipe varying between 0.43 and 1.80 cm. In addition, the results of Iwagwi et al. (2020) indicated 2.6-3.3 cm of variation in the height of the stipe.The size of the pileus ($cm^2$) was also assessed. T1 had the largest size at 18.08 $cm^2$, followed by T4 with 17.33 $cm^2$, T2 with 16.61 $cm^2$ and the smallest value was 15.88 $cm^2$ in T3. Studies have shown that the diameter of the pileus ranges from 4.23 to 14.12 cm (JOHNIE, 2023), which suggests that the values obtained in this study are higher than those in the literature. The size of the pileus is an important characteristic for the commercialisation of products, since it directly influences production factors such as yield.

**Table 5:** Biological aspects of Pleurotus ostreatus var. florida found in the first production flow.

| Treatments | Spawn development | Growth - vegetative period (days) | Total cycle (days) | Stipe size (cm) | Stipe width (cm) | Pileus size (cm )2 |
|---|---|---|---|---|---|---|
| T1 | +++ | 17 | 24 | 1,80±0,35 | 0,76±0,19 | 18,08±0,29 |
| T2 | +++ | 17 | 24 | 1,60±0,51 | 0,90±0,61 | 16,61±0,82 |
| T3 | +++ | 17 | 24 | 1,40±0,11 | 0,70±0,22 | 15,88±0,32 |
| T4 | +++ | 17 | 24 | 1,70±0,81 | 0,63±0,61 | 17,33±0,18 |
| (±)EP | | | | 1,93 | 4,10 | 5,85 |
| CV (%) | | | | 4,38 | 7,94 | 5,94 |

+ sparse growth; ++ moderate growth; +++ abundant growth. **CV:** Coefficient of variation**. SE:** Standard error. **T1:** Green bean pod without the bean (80%) + Malt bagasse (20%); **T2:** Green bean pod without the bean (80%) + Malt bagasse (20%) + yeast (5%); **T3:** Green bean pod without the bean (80%) + Malt bagasse (20%) + hot trub (10%); **T4:** Green bean pod without the bean (80%) + Malt bagasse (20%) + hop trub (10%). Source: Author (2024).

With regard to production aspects, mushroom moisture, yield, biological efficiency, productivity, loss of organic matter and compost consumption in the different formulations were assessed (Table 6). Firstly, mushroom moisture ranged from 81 to 83 per cent, with an average of 82.25 per cent. An important production parameter is yield. Treatment T4 had the highest yield, at 19.5 per cent, while the lowest yield was observed in treatment T1, at 16.83 per cent.

Treatment T2 obtained 17.16 per cent and T3 18.2 per cent. In the research carried out by Iossi et al. (2023), they presented an average yield of 12-21%, with the highest value being 21.39% for the cultivation of Pleurotus ostreatus var. florida in different substrates. Therefore, the results obtained prove to be viable and applicable to the cultivation of this mushroom species in the different treatments based on grain-free bean pods and malt bagasse, supplemented with residual yeast and hot and hop trub.Biological efficiency measures the conversion efficiency of the substrate in the formation of the basidiome, and is an essential factor in verifying the suitability of the medium (RAMAN, 2020). The higher the value of this parameter, the better the suitability of the substrate composition, considering profitability in mushroom production, the values should be higher than 50% (KUMLA et al., 2020). The T4 formulation showed the best biological efficiency with 76.80%, followed by T3 with 72.80%, after T1 with 67.33% and T2 with 66.02%. All the results were greater than 50%, which supports the possibility of commercial use of these formulations.

**Table 6:** Productive aspects of Pleurotus ostreatus var. florida grown in the different treatments.

| Treatments | Mushroom moisture (%) | Yield (%) | Biological efficiency (%) | Productivity (g/day) | Loss of organic matter (%) | Compost consumption (%) |
|---|---|---|---|---|---|---|
| T1 | 82,00±0,19 | 16,83±1,27 | 67,33±0,49 | 0,75±0,22 | 36,66±0,19 | 63,33±1,89 |
| T2 | 81,00±0,58 | 17,16±0,69 | 66,02±0,20 | 0,80±0,59 | 44,87±0,51 | 55,12±0,40 |
| T3 | 83,00±0,28 | 18,2±0,59 | 72,80±1,93 | 0,62±0,61 | 44,80±0,68 | 55,20±0,59 |
| T4 | 83,00±0,50 | 19,5±1,02 | 76,80±0,80 | 0,66±0,49 | 37,60±0,41 | 62,40±0,97 |
| Standard deviation | 3,91 | 7,18 | 1,95 | 3,89 | 2,90 | 4,75 |
| CV (%)§ | 9,83 | 18,76 | 20,65 | 23,91 | 7,59 | 15,30 |

Untransformed data. § Coefficient of variation. Source: Author (2024).

Biological efficiency and productivity are related to the ability to biotransform waste into basidiomata, i.e. the formation of the fruiting body. It is known that the variation in these values is influenced by the type and chemical composition of the substrates used (ZÁRATE-SALAZAR et al., 2020). With regard to productivity, which is an extremely important factor in verifying the viability and commercial sustainability of substrates, it is clear that treatment T2 obtained the best result, with 0.80 g/day, which indicates that this formulation provides the fastest bioconversion of the waste used, while T3 had the worst performance, with 0.62 g/day. Treatments T1 and T4 had yields of 0.75 g/day and 0.66 g/day, respectively.The loss of organic matter is a parameter that dictates whether the formulation is rich in essential nutrients for the fungi; the higher its value, the faster the fungi grow. colonisation and degradation of the substrate, thus reducing the time needed for fruiting (DE PAIVA et al. 2021). In this experiment, the loss values ranged from 36.66% to 44.87%. The highest losses were observed in treatments T2 and T3, with similar values of 44.87% and 44.80%, respectively, followed by treatment T4 with 37.60% and T1 with 36.66%. These data show that the lower the loss of PMO, the greater the possibility of obtaining a second production flow, since the substrates still have nutrients that can be utilised by the fungal metabolism.As for analysing compost consumption, this result shows the support of the substrates in guaranteeing the availability of macro and micro nutrients essential for vegetative development and basidiom formation. Therefore, T1 (63.33%) and T4 (62.40%) absorbed more nutrients than T2 (55.12%) and T3 (55.50%). PMO and CC are complementary results. Together they show both the efficiency of the substrate in ensuring vegetative and reproductive development and the quality and size of the basidiom formed (TAVARWISA et al., 2021). The macroscopic structure formed (fruiting body) in these substrates shows potential for commercialisation, as illustrated in Figure 18. The fruiting bodies have the characteristic whitish to cream colour of the species (JARIAL, 2020). It can also be seen that fruiting was abundant in all the formulations tested, with the presence of voluminous pileus, short and thick stipes, as well as the visualisation of the hymenium formed by lamellae. These macroscopic aspects indicate that development occurred as expected, making all the treatments viable for cultivating Pleurotus ostreatus var. florida.

**Figure 18:** Fruiting of Pleurotus ostreatus var. florida in the different treatments.

**T1:** Green bean pods without the grain (80%) + malt bagasse (20%); **T2:** Green bean pods without the grain (80%) + malt bagasse (20%) + residual yeast (5%); **T3:** Green bean pods without the grain (80%) + malt bagasse (20%) + hot trub (10%); **T4:** Green bean pods without the grain (80%) + malt bagasse (20%) + hop trub (10%). Source: Author (2024).

### 5.5. Correlation of biological and production parameters of Pleurotus ostreatus var.florida in terms of carbon (C), nitrogen (N) and C:N ratio.

As shown in the table of production parameters (Table 6), the treatment supplemented with hop trub (T4) had the best performance in terms of yield and biological efficiency. This factor may be related to the chemical composition of the trub lüpulo, prov v 1m nt riomibrsrsinsmrgs (a- B-acids) from hops (DURALOO; SILVA; BOGUSZ, 2019), humidified with beer and the higher C:N ratio (Table 4), and this formulation may have achieved a better balance. According to Bellettini et al. (2019), the carbon-nitrogen ratio should be within the range of 25-100/1, but depending on the species, different values are needed to achieve the ideal balance of this ratio (RITOTA et al. 2019; ALBUQUERQUE; SOUSA, 2021). T1 and T2 showed the lowest results in terms of yield and biological efficiency, as they were the formulations with the highest levels of nitrogen (2.40 per cent). Therefore, these results indicate that the high level of nitrogen had a negative effect on the production of Pleurotus ostreatus var. florida. It has already been proven that high levels of nitrogen can compromise lignin degradation because they inhibit lignolytic enzymes, making it impossible for the mycelium to degrade lignin and absorb essential nutrients (BELLETTINI et al. 2019).According to Muswati et al. (2021), the decrease in carbon and nitrogen in the process of ruti, sta ir t m nt r 1 ion à isponibili C : N coming from the lignocellulose present in the substrate, since a lower ratio favours the formation of basidiomata. However, in the results presented in this experiment, the highest C:N ratio obtained had the best production results, which may indicate that the 31.9:1 ratio presented by T4 is in the ideal range for the production of this fungus.Another aspect to consider is that the treatment without supplementation of trub and yeast (T1) had the lowest result in terms of loss of organic material, which suggests that this formulation was the least utilised by the fungus, which may indicate a lower amount of essential nutrients for production.Although all the formulations had promising results in terms of production parameters, T4 proved to be the most effective formulation, with the highest yield and best biological efficiency, i.e. the composition of bean pods, malt bagasse added to hop trub achieved an ideal balance in the C:N ratio for the axenic cultivation of P. ostreatus var florida, making it ideal for maximising this cultivation.

## 6. CONCLUSIONS

Based on the results obtained under the conditions established and evaluated in this study, it can be concluded that:

- It was possible to complete the cultivation (vegetative and reproductive phase) of Pleurotus ostreatus var. florida in a formulation based on grain-free bean pods and malt bagasse supplemented with residual yeast and hot and hop trub;

- The biological and production parameters assessed in the 4 treatments show that P. ostreatus var. florida has a short reproductive cycle (24 days) and the EB% proved that the substrates were suitable for growing this species;

- All treatments showed a C:N ratio within the range considered good for the vegetative and productive development of P. ostreatus var. florida;

- The formulation with hop trub supplementation (T4) proved to be the most efficient in terms of production parameters.

However, regardless of the treatment used (T1, T2, T3 and T4), all the brewery by-products used and the bean pods were efficient in the growth and fruiting of the P. ostreatus var. florida species, showing the versatility of these by-products, which can be easily substituted in the event of unavailability.

## REFERENCES


AHMED, Adel F.; MAHMOUD, GAE; HEFZY, M.; LIU, Z.; & MA, C. Overview on the edible mushrooms in Egypt. **Journal of Future Foods**, 2023, 3.1: 8-15.

ALBUQUERQUE, V.; SOUSA, A.C.B. MUSHROOM FUNGICULTURE Pleurotus ostreatus var. florida. 1.ed. **Editora Novas Edições Acadêmicas,** p. 69. 2021.

AMORIELLO, Tiziana; CICCORITTI, Roberto. Sustainability: Recovery and reuse of brewing-derived by-products. **Sustainability**, v. 13, n. 4, p. 2355, 2021.

BELLETTINI, M. B.; FIORDA, F.A.; MAIEVES, H.A. et al. Factors affecting mushroom Pleurotus spp. **Saudi Journal of Biological Sciences**, v. 26, n. 4, p. 633-646, 2019.

BERNARDI, E.; NASCIMENTO, J. S. Cultivation of Pleurotus sajor-caju in different pasteurised substrates. Archives of the Biological Institute, v. 78, p. 217-223, 2020.

BONATTI, M. et al. Evaluation of Pleurotus ostreatus and Pleurotus sajor-caju nutritional characteristics when cultivated in different lignocellulosic wastes. **Food Chemistry**, v. 88, n. 3, p. 425-428, 2004.

COELHO NETO, DM et al. Chemical Concepts Involved in Beer Production: A Review**. Revista Virtual de Química**, v. 1, p. 1-28, 2020.

COÊLHO, J. D. Grain production: beans, corn and soya. ETENE Sector Notebook (Technical Office for Economic Studies in the Northeast). Banco do Nordeste. 2019.

DE GIACOMETI, MARJORIE. In vitro activity of Pleurotus ostreatus, Pleurotus florida and Pleurotus djamor on cyathostomine eggs in horses. 2020.

DE PAIVA, GISEUDO Aparecido et al. Production of edible mushrooms from agro-industrial waste in the Mato Grosso Amazon. **Research, Society and Development**, v. 10, n. 14, p. e548101422523-e548101422523, 2021.

DOS REIS JUNIOR, EIDILSON SOARES et al. Chemical Composition, Antioxidant and Antiproliferative Effect of the Brazilian Mushroom Pleurotus Ostreatoroseus Singer. **Brazilian Journal of Development**, v. 7, n. 4, p. 34642-34656, 2021.

DOS SANTOS GIULIANI, C.; ALVES, A. F.; CIROLINI, A., & DA ROSA, V. P.Preparation of fibre-enriched cake using brewery waste**. Brazilian Journal of Development**, v. 5, n. 9, p. 15396-15404, 2019.

DURELLO, Renato S.; SILVA, Lucas M.; BOGUSZ, Stanislau. Chemistry of hops. **Química Nova**, v. 42, p. 900-919, 2019.

EGÍDIO, K.; CARLOS, B.; CARMO, S.; MELO, R.; MORAIS, M. N. Obtaining Sugar
Fermentable by Acid Hydrolysis of Bean Pods. **In: Brazilian Congress of Chemical Engineering in Scientific Initiation**, 2019.
FERREIRA, DAVID GARCIA; DE ALMEIDA NETO, ONOFRE BARROCA; ASSUNÇÃO, LAÉLIA SOARES. Using the Jun-Cao Technique to Grow Edible Mushrooms in Brazil. **Caderno de Propriedade Intelectual e Transferência de Tecnologia,** v. 2, n. 1, p. 52-63, 2020.
FURLAN, S. A. et al. Mushroom strains able to grow at high temperatures and low pH values. **World Journal of Microbiology and Biotechnology**, Vol. 13, p. 689-692, 1997.
GALAPPATHTHI, M. C. A. et al. Nutritional and medicinal benefits of oyster (Pleurotus) mushrooms: a review. **Fungal Biotec,** v. 1, n. 2, p. 65-87, 2021.
Gomes, F. P. Curso de estatística experimental. Piracicaba: Nobel, ed.13, 1990, p. 476, 1990. GONZÁLEZ, A. et al. Edible mushrooms as a novel protein source for functional foods.
**Food & function**, v. 11, n. 9, p. 7400-7414, 2020.
GUPTA, S.; SUMMUNA, B., GUPTA, M.; & ANNEPU, S. K.Edible mushrooms:cultivation, bioactive molecules, and health benefits**. Bioactive molecules in food**, v. 1, p. 1- 33, 2018.
HOLTZ, M. et al. cultivation of Pleurotus ostreatus using cotton waste from the textile industry**. Journal of Environmental Sciences,** v. 3, n. 1, p. 37-51, 2009.
HU, Yanru et al. Effects and Mechanism of the Mycelial Culture Temperature on the Growth and Development of Pleurotus ostreatus (Jacq.) P. Kumm. **Horticulturae,** v. 9, n. 1, p. 95, 2023.
IOSSI, Matheus R. et al. Macromineral consumption by Pleurotus ostreatus var. florida in different straws. **Archives of Biological Sciences,** n. 00, p. 2-2, 2023.
IBGE - BRAZILIAN INSTITUTE OF GEOGRAPHY AND STATISTICS. **Production of beans in Brazil.** Brazil, 2022.

IBOPE INTELLIGENCE. **Public Opinion Poll on Vegetarianism**. Brazil, 2021.
IWUAGWU, Mary O.; NWAUKWA, Deborah S.; NWARU, Chidi E. Use of Different Agro- Wastes in the Cultivation of Pleurotus ostreatus (Jacq.) Kummer. **Journal of Bioresource Management,** v. 7, n. 2, p. 4, 2020.
JARIAL, R. S. et al. Comprehensive review on oyster mushroom species (Agaricomycetes): Morphology, nutrition, cultivation and future aspects. **Heliyon**, 2024.
JOHNNIE, Christina; SEECHARRAN, Diana; ANSARI, Abdullah Adil.

Comparative yield, yield related parameters and elemental composition of oyster mushroom (Pleurotus ostreatus)grown on different substrates: Mushroom culture. **Mushroom Research**, v. 32, n. 1, p. 27-33, 2023.
KARLOVIé, An r t 1. By-products in the malting and brewing industries-re-usage possibilities. **Fermentation**, v. 6, n. 3, p. 82, 2020.
KOTOWSKI, Marcin Andrzej. History of mushroom consumption and its impact on traditional view on mycobiota-an example from Poland. **Microbial Biosystems,** v. 4, n. 3, p. 1-13, 2019.
KRUPODOROVA, T. et al. Strain-specific features of Pleurotus ostreatus growth in vitro and some of its biological activities. **BMC biotechnology**, v. 24, n. 1, p. 9, 2024.
KUMLA, Jaturong et al. Cultivation of mushrooms and their lignocellulolytic enzyme production through the utilisation of agro-industrial waste. **Molecules**, v. 25, n. 12, p. 2811, 2020.
LACERDA, Sthefany. Edible mushrooms: health benefits, market and aspects of their production. **Agron Food Academy.** UNESP Botucatu Campus São Paulo, p. 1-4, 2021.
LESA, Kaisun Nesa et 1. Nutrition 1 V 1u , M i in 1 Import n , n H 1th-Promoting Effects of Dietary Mushroom (Pleurotus ostreatus)**. Journal of Food Quality**, v. 2022, n. 1,p. 2454180, 2022.
LOMUSCIO, ELISABETTA et al. Durum wheat fresh pasta fortification with trub, a beer industry by-product. **Foods**, v. 11, n. 16, p. 2496, 2022.
MAJESTY, Duru et al. Nutritional, anti-nutritional and biochemical studies on the oyster mushroom, Pleurotus ostreatus. **EC Nutrition**, v. 14, n. 1, p. 36-59, 2019.
MANAN, Sehrish et al. Synthesis and applications of fungal mycelium-based advanced functional materials. **Journal of Bioresources and Bioproducts**, v. 6, n. 1, p. 1-10, 2021.
MARADINI FILHO, Antonio Manoel et al. Utilisation of agro-industrial waste. **Vice-Rector,** v. 29075, p. 287, 2020
MASSARDI, MATHEUS MOREIRA; MASSINI, RODRIGO MANHÃES MOTTA; DE JESUS SILVA, DEUSANILDE. Chemical characterisation of malt bagasse and evaluation of its potential for obtaining value-added products**. The Journal of Engineering and Exact Sciences**, v. 6, n. 1, p. 0083-0091, 2020.
MATHIAS, T. R. S.; MELLO, P. P. M; SERVULO, E. F. C. Characterisation of brewery waste. **In: Brazilian Congress of Chemical Engineering**. 2014. p. 1-8.
MLECZEK, MIROSLAW et al. Changes in mineral composition of six strains

of Pleurotus after substrate modifications with different share of nitrogen forms. **European Food Research and Technology**, v. 247, p. 245-257, 2021.

MORETTO, E; FETT, R; GONZAGA, L.V; KUSKOSKI, E.M. 2002a. Composition centesimal of plants and other materials. 2nd ed. UFRGS, 174 p. **Food products**. Chapter 1. in: Moretto et al. (eds). Introduction to food science.

MUSWATI, C.; SIMANGO, K.; TAPFUMANEYI, L.; MUTETWA, M.; NGEZIMANA, W.The effects of different substrate combinations on growth and yield of oyster mushroom (Pleurotus ostreatus). **International Journal of Agronomy**, vol 21, p-10, 2021.

NINDJIO, GAËL FERDINAND KAZÉ et al. Lignocellulosic-based materials from bean and pistachio pod wastes for dye-contaminated water treatment: optimization and modelling of indigo carmine sorption. **Polymers**, v. 14, n. 18, p. 3776, 2022.

OKERE, S. et al. Effects of different carbon/nitrogen ratios on yield and bioconversion efficiency of Pleurotus ostreatus cultivated on cassava peels and sawdust-based substrate. **Ajol. Info**, v. 34, n. 1, 2021.

PAZZA, Aline Carina Vieira et al. Nutritional composition and physiological functional properties of edible mushrooms. **FAG JOURNAL OF HEALTH (FJH),** v. 1, n. 3, p. 240-265, 2019.

RAJARATHNAM, S.; BANO, Z.; STEINKRAUS, K. H. Pleurotus mushrooms. Part III. Biotransformations of natural lignocellulosic wastes: commercial applications and implications. **Critical Reviews in Food Science & Nutrition**, v. 28, n. 1, p. 31-113, 1989.

RAMAN, JEGADEESH et al. Cultivation and nutritional value of prominent Pleurotus spp.: an overview. **Mycobiology,** v. 49, n. 1, p. 1-14, 2021.

RATHOD, M.; GADADE, R. B., THAKUR, G. M. et al. Oyster mushroom: cultivation, bioactive significance and commercial status**. Frontiers in Life Science**, v. 2, p. 21, 2021.

DOS REIS JUNIOR, Eidilson Soares et al. Chemical Composition, Antioxidant and Antiproliferative Effect of the Brazilian Mushroom Pleurotus Ostreatoroseus Singer. **Brazilian Journal of Development**, v. 7, n. 4, p. 34642-34656, 2021.

RITOTA, M.; MANZI, P. Pleurotus spp. cultivation on different agri-food by-products: Example of biotechnological application. **Sustainability**, v. 11, n. 18, p. 5049, 2019.

SILVA, MAFALDA et al. Pre-and Postharvest Strategies for Pleurotus ostreatus Mushroom in a Circular Economy Approach. **Foods,** v. 13, n. 10, p. 1464, 2024.

SOCCOL, CARLOS RICARDO et al. Mushrooms: a promising source of active compounds for the development of pharmaceutical and nutraceutical products.

In: **Biotecnologia Aplicada à Agro&Indústria-Vol**. 4. Blucher Open Access, 2017. p. 315-360.
STEFFEN, G.P.K; STEFFEN, R. B.; HANDTE, V.G. Edible mushroom production in Brazil: a growing market, 2019.
TAVARWISA, M., GOVERA, C., MUTETWA, M. NGEZIMANA, W. Evaluating the suitability of baobab fruit shells as substrate for growing Oyster Mushroom (Pleurotus ostreatus). **International Journal of Agronomy**. v.7, p.6620686, 2021.

TASKIRAWATI, Ira et al. The bamboo sawdust and addition of em4 as an alternative material for the cultivation of oyster mushroom (Pleurotus ostreatus). In: **IOP Conference Series: Earth and Environmental Science**. IOP Publishing, 2020.
TEDESCO, M.J; GIANELLO, C; BISSANI, C.A; BOHNEN, H; VOLKWEISS, S.J. 1995.
Soil analyses, UFSC, p. 19-56.
TORTORA, GERARD J.; CASE, CHRISTINE L.; FUNKE, BERDELL R. **Microbiology-12th Edition.** Artmed Editora, 2016.
URBEN, A. F., & URBEN, A. F. Mushroom production using modified Chinese technology: biotechnology and applications in agriculture and health. **EMBRAPA**, 2017.
VAZ JUNIOR, SILVIO. Utilising agro-industrial waste: a sustainable approach. **EMBRAPA**, 2020.
ZAKIL, A.F; ISA, M.R; SUE, M.S.M; ISHA, R. Growth performance and mineral analysis of Pleurotus ostreatus (oyster mushroom) cultivated on spent mushroom medium mixed with rubber tree sawdust. **Materials Today: Proceedings**. vol. 57, p. 1329-1337, 2022.
ZÁRATE-SALAZAR, JHONATAN RAFAEL et al. Use of lignocellulosic corn and rice wastes as substrates for oyster mushroom (Pleurotus ostreatus Jacq.) cultivation. **SN applied Sciences**, v. 2, p. 1-10, 2020.
ZENG, XU et al. Nutritional characterization and untargeted metabolomics of oyster mushroom produced using astragalus membranaceus var. Mongolicus stems and leaves as substrates. **Frontiers in Plant Science**, v. 13, p. 802801, 2022.
ZHENG, Y; YICHUN, X; XIE, Y; YU, S. Asexual reproduction and vegetative growth of Bionectria ochroleuca in response to temperature and photoperiod. **Ecology and Evolution**. Vol. 11, p. 10515-10525, 2021.

MIX
Papier aus verantwortungsvollen Quellen
Paper from responsible sources
FSC® C105338

Printed by Books on Demand GmbH, Norderstedt / Germany